Norfolk

25 Years of the Wherry Lines

CHRIS BOON

BRITAIN'S RAILWAYS SERIES, VOLUME 20

Front and back cover images: see page 69 for details.

Title page image: see page 50 for details.

All images, unless specified, belong to the author.

Published by Key Books
An imprint of Key Publishing Ltd
PO Box 100
Stamford
Lincs PE19 1XQ

www.keypublishing.com

ISBN 978 1 80282 027 0

Typeset by SJmagic DESIGN SERVICES, India.

Introduction

The Norfolk resort of Great Yarmouth was, for many years, a popular destination for enthusiasts and photographers, owing to the wide range of loco-hauled services, in particular on summer Saturdays, delivering coachloads of holidaymakers for their week in the sun. Through services operated to Yarmouth from a wide range of destinations, including Newcastle, Birmingham, Nottingham, and Walsall, as well as the extension of InterCity workings between London Liverpool Street and Norwich. These services provided an opportunity to ride behind a range of locomotives, with unusual freight engines often deployed. The cross-country services via Peterborough and Ely operated until 1993, while the direct trains to the capital continued until 2014.

In addition to these long-distance loco-hauled services, diesel traction has also been used on local trains from Norwich, usually to strengthen the often-stretched fleet of diesel multiple units or for additional capacity for special events. While ad-hoc loco haulage could be encountered on these local services for many years, the summer of 1994 produced a regular pattern with Class 37/0s appearing every Saturday between July and September. This operation continued in 1995 and 1996, with traction changing to Class 31/4s in mid-1995. In more recent years, these local services became known as the 'short set' to distinguish them from the full rakes of stock used on long-distance services.

After 1996, the sight of loco haulage on the branches became rather less predictable, other than on the summer-dated holidaymaker trains from the capital, often known as the 'drags', owing to the regular Liverpool Street to Norwich sets being dragged to the coast by a Class 47. These Brush Type 4s, and occasionally more exotic traction, did still periodically appear, covering for the regular DMUs on the lines to Great Yarmouth and Lowestoft. These became known as the Wherry Lines, after the cargo boats once seen on rivers in Norfolk and Suffolk.

From 2000, the annual Lowestoft Airshow brought a range of unusual traction to the most easterly town in Britain, with the hugely popular two-day event requiring significantly more capacity to bring visitors to the seafront air festival. The first year heralded the return of Class 31/4s to the region, by now hired from Fragonset, with EWS Class 67s and a '47' in 2001, which at the time could be found in Norwich for the daily postal services. The traction choices settled into a regular pattern of Class 47s in subsequent years, although examples from Freightliner, Cotswold Rail, Advenza Freight and Direct Rail Services all featured. Despite attracting around 100,000 visitors each year, the airshow struggled to cover its costs, with the last event taking place in 2012, when DRS' newly refurbished Class 37/4s made an appearance, much to the delight of the enthusiast community.

Another event that brought even more exotic traction to East Anglia was the Association of Community Rail Partnerships (ACoRP) weekend in September 2005, which featured a loco display in Norwich, plus additional services on the Wherry Lines hauled by Classes 20, 33 and 47. It had been hoped that this would be an annual event, with similar operations in different locations each year, but sadly it proved to be a one-off. Other occasions which required the augmentation of the usual diesel units included Pop Beach, a one-day music event in Great Yarmouth, and the one-off airshow also in Great Yarmouth in 2018.

From 2010, loco haulage started to appear again on the Wherry Lines on a more frequent basis, with traction now provided by DRS instead of Cotswold Rail. Other than rare appearances of a Class 20/3 and Class 57/0, the featured traction initially came from the DRS Class 47 fleet before changing to

Class 37/4s in 2015, to the excitement of many. These services settled into a regular pattern with a daily weekday diagram, initially utilising the hauled set if no unit was available, before becoming booked daily services and eventually extending to Saturdays as well. DRS' newest fleet, the Class 68s, also made an appearance in 2016–17, on a second set that was deployed to cover for a Class 170 Turbostar unit undergoing collision damage repairs. This brought the opportunity to see regular loco haulage from two different classes on the Norfolk branches.

With the 2016 franchise renewal, a fleet of bi-mode Flirt units was ordered from the Swiss manufacturer, Stadler, to be deployed on all branches in Norfolk and Suffolk. As these new units gradually came into service, it was clear that the days of regular loco haulage were numbered, especially as the 1970s Mk.2 coaches in use then did not conform to the accessibility standards that came into force from 1 January 2020. As well as organising three charters to celebrate these loco-hauled services, Greater Anglia also announced the final day of Class 37 operation in advance, with an excellent send-off taking place on 21 September 2019.

This book celebrates the 'short set' local services along with those for special events, notably the Lowestoft Airshow, in the 25 years from 1994 to 2019. This book is dedicated to my parents, Brenda and Peter, for patiently waiting on cold stations in the early years while I took my photos, and to my endlessly patient wife, Sue, for looking after our son, Tom, while I dashed out to ride on every possible Class 37 and 68, and for tolerating many family outings carefully planned around the operation of the 'short set'.

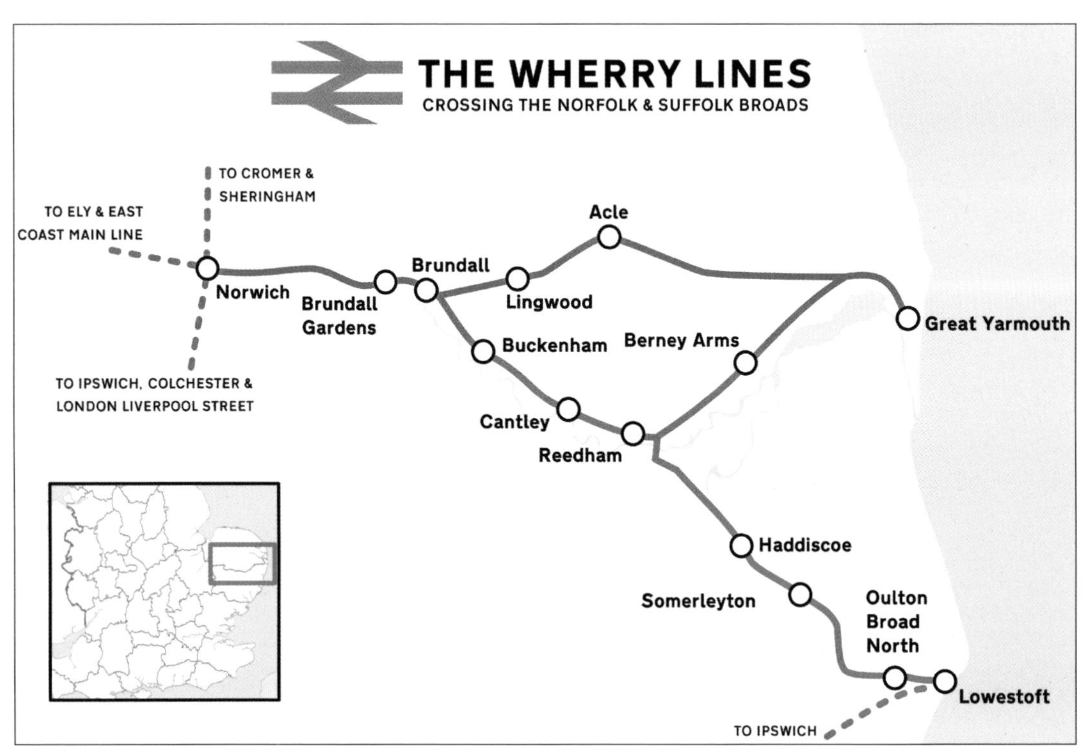

Wherry Lines map (Rcsprinter123, CC BY 3.0, Wikipedia)

Norfolk Rail: 25 Years of the Wherry Lines

With the Class 47-hauled cross-country services to Great Yarmouth having finished the year before, 1994 could have been an unexciting year. However, in addition to the traditional Class 47-hauled services from Liverpool Street to Yarmouth, 37077 was used on 9 July for two return trips from Norwich to Yarmouth, plus one return to Lowestoft, top and tailed with 47476. A similar pattern emerged each week until 3 September, with a range of Class 37/0s working a short set of stock, mostly on Yarmouth services. These workings could be seen as the origins of the East Anglian 'short set'. On 13 August, the chosen traction was 37167, which is seen here arriving in Norwich with the 11.19 from Great Yarmouth. The uniquely painted Class 37 went on to work another two round trips to Yarmouth, plus two returns to Lowestoft, which would have required the stock to be propelled into the yard, owing to the absence of run-round facilities at the Suffolk resort. 47575 *City of Hereford* can be seen on the right, having worked the 10.43 from Yarmouth.

Stratford depot in East London could always be trusted to produce one-off colour schemes. Clearly inspired by the standard triple grey livery, but with different shades, plus the addition of a narrow blue bodyside band, 37167 was definitely unique! This was a relatively short-lived colour scheme, being applied to the Type 3 in July 1994. The spring of 1995 saw 37167 emerge from the paint shop to become the third Type 3 to wear the new Mainline Freight blue livery. The loco is seen running round the stock in Norwich before working the 2P20 12.35 back to Great Yarmouth.

Now preserved on the East Lancashire Railway in BR blue, 37109 wore Trainload Metals colours back in 1994 and took a trip to the seaside on 20 August. The 'Tractor' is seen at Norwich before working 2P20, the 12.35 to Great Yarmouth. Note the Mk.1 buffet behind the loco, which was a popular addition to the 'short set' on many weeks. The last Mk.1 buffets remained in service with Anglia Railways until 2000, being some of the last passenger-carrying Mk.1 vehicles in regular main line service.

Metals-branded 37109 is pictured running round the stock in Great Yarmouth after working 2P20, the 12.35 from Norwich. On this day, the attractive Type 3 worked three returns between Norwich and Yarmouth. 37109 could usually be found employed on engineering trains in East Anglia, being part of the Stratford-based Anglia Infrastructure pool. The station roof once extended over the ends of the platforms but was removed in 1982. However, the concourse still enjoys protection from the elements, unlike neighbouring Lowestoft, where the whole roof was removed in 1992.

On the penultimate week of the season, a rare moment of sunshine illuminates split-box 37013 at Norwich on 27 August before working 2P20, the 12.35 service to Great Yarmouth. The Type 3 was subsequently repainted in the attractive Mainline Freight blue livery but was stored in October 1999 and cut-up at EMR Kingsbury in February 2007. Note the first-class Mk.2 coaches behind the loco; at this time, the 'short set' was formed of available stock from Crown Point depot, this often featuring first class or buffet vehicles.

Triple grey 37013 is pictured being uncoupled from the stock at Great Yarmouth after working 2P20, the 12.35 from Norwich. The loco would work one more round trip to the Norfolk resort on this day. Note the Class 158 DMU in platform 1 on the right; these units had replaced the Class 47-hauled services to the Midlands and the North, although these through trains would only continue for a few more years, subsequently requiring passengers to change in Norwich to reach the coast.

The last Saturday in 1994 to enjoy Class 37-hauled services to Great Yarmouth was 3 September, with another Trainload Metals loco being chosen for the familiar three returns to the seaside. The Anglia Infrastructure pool supplied 37241, which is pictured at Norwich in the yard before shunting to the station to work 2P12, the 09.35 to Great Yarmouth. The last service of the day, 2P25 16.09 Yarmouth–Norwich, would be the last recorded passenger service for this loco, with withdrawal coming in 1996, although scrapping did not occur until 2001 at TJ Thomson's yard in Stockton.

The only chance to experience loco haulage to the Norfolk coast after the final summer Saturday working in 1994 was through the occasional replacement of the usual DMUs by Class 47s and coaching stock, for example on two Saturdays in November where 47567 and 47519 were used to augment busy shoppers' trains to the city. However, the more predictable weekly summer Saturday services continued in 1995, with the season starting slightly later and, featuring refurbished 37676 on 12 August, hauling the usual three returns between Norwich and Great Yarmouth. On 19 August, split-box 37012 worked the familiar diagram, being pictured here departing from Brundall with 2P18, the 11.37 Norwich–Great Yarmouth. This day witnessed the last Class 37-hauled scheduled passenger service on the Wherry Lines until the DRS pair of 37419 and 37425 were used on the 2012 Lowestoft Airshow specials, and then 37405 and 37425 took over from Class 47s on the Greater Anglia 'short set' on 17 June 2015, some 20 years later.

After two weeks of Class 37 haulage, the three Saturday return trips to Great Yarmouth during summer 1995 transferred to Class 31s. The Class 31/4s offered the advantage of being able to supply power for the air-conditioned Mk.2 coaches. Rail blue 31459, from the Derby-based infrastructure pool, pulls away from Norwich on the 14.35 to Great Yarmouth on 26 August. Class 31s had been a familiar sight in Norwich for many years, working services to Birmingham, until May 1988 when they were replaced by Class 156 DMUs.

The first loco-hauled 'local' of the day on 9 September 1995 was the 09.35 departure from Norwich. On this day, the chosen traction was InterCity Mainline 31407, again from the Derby infrastructure pool. The ensemble is pictured pausing at Brundall before heading east towards the Norfolk coast. The area around Brundall station has become busy with boatyards, with the level crossing providing access to these from the main village located north of the railway.

On the same day, the final few customers board the 11.37 departure from Norwich for Great Yarmouth with 31407 at the helm, providing the unusual sight of a matching loco and stock. 31407 also worked these services in 1996, but had been repainted into Mainline Freight livery by then. Despite having a reputation for being underpowered, the 1,470hp English Electric power unit in 31407 would have been perfectly adequate to haul the five Mk.2 coaches to the coast.

By summer 1996, the Electric Train Heat (ETH)-fitted Class 31s based in Derby were owned by Mainline Freight, one of the pre-privatisation shadow freight companies. The Type 2s operated the familiar three returns from Norwich to the coast. On 17 August, 'Dutch'-liveried 31466 was the chosen motive power, being illustrated at Brundall where the driver is checking the guard's signal to depart, while working the 09.35 Norwich–Great Yarmouth. 31466 also worked to Yarmouth on 24 August, being entrusted with the 1P51 10.43 Great Yarmouth to Liverpool Street as far as Norwich, and the 10.55 Liverpool Street to Yarmouth back to the coast, both services featuring a full 10-coach InterCity set! This loco was later to become the only member of its class to wear EWS livery in service, and is now preserved at the Dean Forest Railway.

Bringing back memories of the Norwich–Birmingham services in the 1980s, uniquely liveried Mainline Freight 31407 is illustrated at Norwich on 31 August 1996, with the 11.37 departure for Great Yarmouth. Note the pointwork under the stock, which would have been utilised for the run-round manoeuvre after the first trip of the day to the coast. The current Norwich station, originally one of three in the city, was opened as Norwich Thorpe in 1886, replacing the previous terminus which was, by this stage, inadequate to handle the volume of rail traffic.

The colourful combination of 31407 and the InterCity stock, complete with DBSO brake van, waits at Great Yarmouth with the 12.06 back to Norwich on 31 August. Rapid run-round manoeuvres at the coast were required to keep to the timetables designed for DMU operation. After a period of storage at Barrow Hill, 31407 was scrapped by Ron Hull Jr in Rotherham in October 2006.

A week later, on 7 September, the aircraft blue loco could again be found employed on Norwich to Great Yarmouth services, making an attractive sight at Acle while working the 11.37 from Norwich to the Norfolk seaside. When the line from Brundall to Breydon Junction opened in 1883, Acle village was a convenient location to provide a passing loop on the otherwise single line. The only other station on this section is Lingwood, between Acle and Brundall.

After the summer of 1996, the use of additional stock to supplement the usual diesel units on the Anglia branches became more sporadic. Although not a 'short set', the preserved Hastings DEMU provided additional capacity for Anglia Railways and interest for enthusiasts, with regular use in 1998–99, as illustrated at Oulton Broad North on 29 August 1998 where 1001 was working a Norwich–Lowestoft service. Various other additional units were hired by East Anglia TOCs in subsequent years, including First North Western's 153313 in 2007 and Arriva Trains Wales' 150280 and London Midland's 153354 and 153364 in 2011.

In the late 1990s when the 'short set' did run, it made use of available traction in the Norwich area, with local 'Thunderbird' loco 47825 *Thomas Telford* working to Great Yarmouth on 4 June 1999 and 50050 *Fearless* having a trip to the coast on 10 July, while based at Crown Point. In August 2000, the 'short set' returned to provide additional capacity for the Lowestoft Airshow, which took place on 3–4 August. By now, privatisation had enabled various companies to purchase ex-BR locos for 'spot hire' and smaller contracts, the most significant being Fragonset Railways, which boasted a fleet of Class 31s and 47s. Two members of the former class were hired by Anglia Railways for the airshow services, with 31468 *Hydra* pictured leading 31459 *Cerberus* and passing a rake of OBA wagons in Lowestoft yard, taking the empty stock back to Norwich after working the 09.23 to Lowestoft.

The next service to bring visitors to the Lowestoft Airshow was the 11.26 from Norwich, pictured crossing Reedham swing bridge, with 31459 *Cerberus* leading 31468 *Hydra*. Note the InterCity Mk.1 buffet in the formation in its final year of service, sandwiched between Anglia Railways Mk.2s. The air display took place by Lowestoft seafront during the afternoon, with the Class 31s then taking visitors back to Norwich at 17.53. The airshow became a popular event for enthusiasts over the following 12 years, with a wide range of traction being employed on these additional services.

After the 2000 Lowestoft Airshow, Anglia Railways continued to use Fragonset Class 31s on additional services to Lowestoft and Great Yarmouth for a few weeks, with workings noted on 19 and 26 August. 31601 *Bletchley Park Station X* is pictured in Norwich preparing to work the 12.58 to Lowestoft on 26 August 2000. The Class 31/6s had through ETH cables but could not provide power for heating and air conditioning themselves, the through wiring being provided for use when the locos double-headed with an ETH-fitted Class 31/4. The new order can be seen either side of the loco hauled set, with an Anglia Railways Class 170 Turbostar on the left and a Central Trains unit on the right.

Showing the advantage of top and tailed operation, partner 31459 *Cerberus* pauses at Lowestoft ready to return to Norwich on 26 August. The Mk.1 buffet had by then been removed from the rake, with a uniform set of Anglia Railways Mk.2s, including a DBSO brake van, being deployed. Lowestoft station has benefitted from many improvements in recent years, with the Lowestoft Central Project opening the former parcels office as an exhibition space, replacing the large wooden doors to open the concourse onto Station Square, and installing heritage lighting.

Class 47 traction was used by Anglia Railways during 2001, with Freightliner's 47292 and Fragonset's 47488 completing seven returns to Lowestoft and Great Yarmouth on Saturday 21 April. On 19 May, Freightliner-owned 47295 and EWS' 47727 *Duke of Edinburgh's Award* were hired to cover a DMU diagram, again completing seven returns to the coast. 47295 is pictured at Great Yarmouth preparing to work 2P33, the 16.21 to Norwich. The following working, the 18.10 from Great Yarmouth, was the last passenger working for the work-stained Type 4 before withdrawal. However, 47727 was still in main line service in 2021 with GB Railfreight, usually seeing employment on stock moves.

The summer 2001 timetable only featured one Class 47-hauled service from Great Yarmouth, with the 10.05 to Liverpool Street booked for Type 4 haulage. However, Norfolk County Council sponsored the loco to be used for additional services to the seaside on Saturdays. On 14 July, 47793 *Saint Augustine* was the chosen traction, being pictured in Norwich coupling up to a short rake of Anglia Railways Mk.2s before working the 11.18 to Great Yarmouth and 13.35 return. 47793 was preserved at Mangapps Railway Museum in 2007 and repainted into BR blue as 47579; it subsequently gained large BR logos and moved to the Mid Hants Railway.

After much debate and rumour, the chosen traction type for the 2001 Lowestoft Airshow shuttle services were top-and-tailed EWS Class 47s and Class 67s. The current Anglia 'Thunderbird' 47757 *Restitution* worked on both days, while 67020 operated on Thursday 26 July and 67015 on Friday 27 July. Illustrated awaiting departure from Norwich with the 10.25 to Lowestoft on the Friday is 67015. The first service, the 08.35 from Norwich, was cancelled on this day, believed to be because of the lights and air conditioning on the coaches being left running overnight, resulting in flat batteries in the morning!

Later on 27 July, 67015 is seen bringing up the rear of the empty stock back to Norwich after the 10.25 to Lowestoft. 47757 *Restitution* is providing power at the front. 47757 took the airshow visitors back to Norwich at the end of the event, working the 1G73 17.45 Lowestoft–Norwich. Note the longer set of stock compared to the previous year; perhaps not a true 'short set', but nonetheless of great interest as Class 67s have very rarely worked passenger services on this route. The only other occasion being when 67024 replaced failed B1 steam locomotive 61264 on 10 November 2001, with a charter from Liverpool Street.

Shunter power! Although Cotswold Rail 08871 did not haul any passenger services on 1 September 2001, the sight of a Class 08 in Norwich station was unusual, with the pilots usually confined to Crown Point depot. The occasion is the arrival of the stock for the 11.18 local service to Great Yarmouth, which would be powered by 47721 *Saint Bede*. The Type 4 had worked the 10.05 Great Yarmouth to Liverpool Street as far as Norwich, before being deployed on this return to the coast.

After working the 11.18 from Norwich, 47721 *Saint Bede* is pictured running round the stock in Great Yarmouth before heading back to the city with the 12.31 service. Note the signal box behind the loco, labelled as 'Yarmouth Vaux' to distinguish its location since the Norfolk town once boasted three stations in the form of Beach, Vauxhall and Southtown. The carriage sidings can just be seen on the far left; these were installed in 1959 to help with the large amount of holiday traffic to the resort now routed to Vauxhall station following the closure of the Midland and Great Northern Joint station at Yarmouth Beach.

During early 2002, EWS Class 47s were used on an occasional basis to cover for unavailability of DMUs, with two usually stabled in Norwich between operating the overnight postal services. On Saturday 6 April, 47789 *Lindisfarne* completed two returns to Great Yarmouth solo, with one trip to Lowestoft and four to Great Yarmouth on Sunday 7 April, featuring 47789 and 47792 *Saint Cuthbert* top and tailing. Then on Monday 8 April, 47789 was paired with Fragonset's 47701 *Waverley* for seven round trips to the coast. The Fragonset loco is pictured arriving into Norwich on 8 April with 2P21, the 13.21 ex Great Yarmouth.

The two heads out of the front window suggest 47789 *Lindisfarne* is a popular choice of traction for the 2D32 13.57 to Lowestoft on 8 April, pictured departing from Norwich. Anglia Railways' 86221 *BBC Look East* can be seen on the left, preparing to depart for London Liverpool Street. The Class 86s operated on the Great Eastern Main Line until 2004, when they were replaced by Class 90s.

On 9 April, 47701 *Waverley* is seen in Jubilee Carriage Sidings in Norwich after working one trip each to Lowestoft and Great Yarmouth with 47789 *Lindisfarne.* Note the InterCity BSO behind the loco, this being unusual as all other Anglia Railways brake vehicles were DBSOs, fitted with a cab for push-pull operation. Platform 6, in the foreground, was built in 1955 and added valuable capacity to Norwich station.

Again, numerous rumours were circulating before the 2002 Lowestoft Airshow about the chosen traction for the specials. The main stories featured either Fragonset Class 31s or Cotswold Rail's 47714 and a Freightliner Class 47. In the end, the latter materialised. The weather on the Thursday was not ideal and the Anglia Railways 'Thunderbird' 47714 was paired with 47309 from Freightliner. The 10.05 and 11.57 departures from Norwich were advertised as loco-hauled, as was the 17.45 return. The weather on the Friday was much better, when 47197 was the chosen Freightliner loco, again paired with 'Shove-Duff' 47714. Although the morning return working to Norwich, 5G71, was booked as an empty stock move, it ran on both days as a service train, providing an extra opportunity for some rare non-ETH Class 47 haulage. Freightliner's 47197 is pictured powering over Reedham swing bridge and heading back to Norwich on 2 August 2002.

The second working on the Friday, 1G71 11.57 Norwich to Lowestoft, was booked to run in front of a unit-operated stopping service. However, as with the earlier run, these services were reversed. Cotswold Rail-owned 47714 is illustrated passing Lowestoft Yard with 47197 out of sight on the rear. The picture was taken from the Iron Bridge, built to link the residential area of Lowestoft with Commercial Road and the docks, it also providing a popular vantage point for photographers. Sadly, the bridge has since been demolished. During the air display in the afternoon, an RAF Harriet jet crashed into the sea by Lowestoft beach owing to engine failure. Luckily the pilot ejected from the plane, and was brought ashore by the lifeboat, with the spectacle seen by thousands of visitors. Bathers were asked to leave the sea, with the area requiring a security presence for the next few days until the £35 million GR7 jet was retrieved.

By 2003, the Cotswold Rail fleet of Class 47s had grown to four. Therefore, the rumours about possible traction were slightly less exotic, it being clear that the specials would be top-and-tailed by two Cotswold '47s', but which ones? Although the two ETH-fitted locos, 47714 and 47818, had unusually worked the 'short set' on Sunday 13 July, Virgin-liveried 47818 and Cotswold silver 47200 *The Fosse Way* were chosen for the airshow specials, the former providing power for the air conditioning and secondary door locks. Thursday 31 July was the day to be out on the lineside, the weather on the Friday being dull and wet all day. No-heat 47200 passes over Reedham swing bridge with the ECS return to Norwich after working the 1G70 10.05 Norwich–Lowestoft.

Also on 31 July 2003, 47200 *The Fosse* Way brings up the rear of 1G72, the 11.57 from Norwich, as the special heads for Lowestoft. The Berney Arms line to Great Yarmouth can be seen on the left. A fan of sidings could once be found to the right of this picture until at least the early 1960s; although the track is long gone, the gap in the embankment is still present, although heavily overgrown. The junction has since been remodelled and simplified as part of the Wherry Lines resignalling project.

Providing another opportunity to enjoy the same pair of Cotswold Rail Class 47s from the airshow specials, but on a much shorter set of stock, 47200 *The Fosse Way* and ex-Virgin 47818 operated three returns each to Great Yarmouth and Lowestoft on Saturday 29 November. The ensemble is pictured in Lowestoft after working the 08.57 from Norwich. The Fosse Way was a Roman road which linked Exeter and Lincoln via Bath, Cirencester and Leicester.

Still wearing a debranded livery from its former operator, with the outline of the Virgin logos visible, 47818 is seen at the bufferstops in the most easterly station in Great Britain with the 09.44 to Norwich. Note the first-class coach next to the loco, this providing some unexpected luxury for passengers on this dull November day. The grain silo on Commercial Road can be seen behind the first coach; the 14,000 tonne capacity silo is part of the facilities of the Port of Lowestoft.

Almost ready to depart, the last few passengers for Norwich hurry along the platform at Lowestoft ready to enjoy some Class 47 haulage; will they appreciate the experience? Cotswold Rail's 47200 *The Fosse Way* is about to power the 09.44 service with 47818 on the rear providing power for the air conditioning. Note the 'green circle' multiple working connector on the former headcode panel as installed by its previous operator Freightliner.

The driver applies the power on 47200 *The Fosse Way* to pull away from Oulton Broad North while working the 09.44 Lowestoft–Norwich. Even with 47818 on the rear, the 2,750hp Brush engine would have no problem hauling just three coaches! The station was originally named Mutford when first opened in 1847, with a single platform to the west of the level crossing. The present platforms, to the east of the crossing, were provided as part of the doubling of the line in 1901.

By 2004, 47813 had joined the Cotswold fleet, on hire from Porterbrook for the Norwich–London diversions while Ipswich Tunnel was closed, and 47714 and 47818 had been repainted into Anglia Railways and 'One' liveries, respectively. The diagrams for the 2004 airshow specials were linked in with the tunnel 'drags', with the 10.05 from Norwich to Lowestoft returning to work the 12.12 to Liverpool Street via Cambridge, and the 11.47 Liverpool Street to Norwich, later working the 17.56 Lowestoft to Norwich. Much to the disappointment of the 15 photographers standing on the bridge at Reedham on Thursday 29 July awaiting 47813 and 47316 on the 1G70 10.05 from Norwich, 170203 and 170206 arrived. It was later found out that there had been a problem with the coupling on 47813 at Crown Point, and despite every effort to solve the problem, the hauled set had to be substituted for the Turbostars at the last minute. The afternoon diagram was slightly more successful with 47714 illustrated near Oulton Broad North on the rear of the late-running 5G77 15.21 Norwich to Lowestoft. 47316 *Cam Peak* was on the front.

The trip to Reedham on the Friday was a little more successful than the day before, with 47818 arriving with 1G70, the 10.05 Norwich to Lowestoft, again with 47316 *Cam Peak* on the rear. Fortunately, the pair of Class 170 Turbostars working the return ECS from the 09.26 Norwich to Lowestoft just cleared the shot in time! Three-car 170208 is leading two-car 170273 towards Norwich to collect more airshow visitors on 30 July.

The 5G79 11.00 ECS return from Lowestoft was worked as a service train on 30 July, providing the only chance for haulage from 47316 *Cam Peak* over the two days of the airshow. The ensemble is illustrated crossing Reedham swing bridge heading back to Norwich, where 47818 worked 1G49, the 12.12 to London via Cambridge. Like so many of the other structures on the line, Reedham swing bridge dates from the doubling of the line, being constructed in 1904 to replace the original single-track swing bridge.

47818 brings up the rear of the 5G79 11.00 Lowestoft–Norwich at Reedham on 30 July, clearly showing the colourful 'One' livery of the new operator, which had taken over from Anglia Railways on 1 April. Note the distant signal, indicating the route through Reedham Junction, where the line from Lowestoft joins the Berney Arms line, was clear. Out of sight to the right is an abandoned formation linking the Lowestoft and Yarmouth lines; it was closed around 1880 and was never used by timetabled passenger services.

Left: The final working for the 2004 airshow, 1G75 17.56 Lowestoft to Norwich, produced Great Western-liveried 47813, again with 47316 *Cam Peak* on the rear. The location is the footbridge just east of Oulton Broad North station, with Leathes' Ham in the background. The land to the right of the fence once housed a large sleeper depot. The site was so large that a narrow gauge railway was provided, with locations across the LNER network and Eastern Region supplied until closure in 1964.

Below: Cotswold Rail's 47818 is illustrated arriving at Lowestoft, past the magnificent mechanical signals on Saturday 19 March 2005, with 2J76 12.57 Norwich–Lowestoft, covering for an unavailable unit. 47714 is on the rear, still sporting the attractive livery of the previous operator, Anglia Railways. In 2005, Cotswold Rail Class 47s were diagrammed to work a daily service from Great Yarmouth to London, with 1V01 06.22 Yarmouth–Liverpool Street booked for Type 4 haulage to Norwich, and 1V06 17.00 Liverpool Street–Yarmouth from Norwich to the coast but seeing the 'short set' in service at a weekend was much more unusual.

With matching Anglia Railways livery, 47714 looks very smart with the short rake of Mk.2 coaches at Lowestoft on 19 March 2005, preparing to work the 2J79 13.43 to Norwich. 'One'-liveried 47818 is on the rear. On the previous weekend, 47714 had visited the Mid Norfolk Railway as part of its spring diesel gala. After the demise of Cotswold Rail, 47714 was acquired by Harry Needle Railroad Company, and was still in use at the Old Dalby test track in 2021.

47714 and 47818 worked all day on 19 March 2005, with an impressive 12 round trips to Lowestoft and Great Yarmouth. 47714 is pictured at Oulton Broad North working 2J91, the 19.48 Lowestoft–Norwich. The Type 4s hauled one return each to Great Yarmouth and Lowestoft after this. The signal on the Norwich-bound platform can be seen behind the loco, with a sighting board behind the signal arm to aid visibility against the station footbridge, which was directly behind the signal until its removal in January 1974.

In September 2005, the Association of Community Rail Partnerships (ACoRP) worked with 'One' Railway to organise a weekend of special loco-hauled services and a display of rolling stock in Norwich. It had been hoped that this would become an annual event, hosted by different train operating companies each year, but sadly this proved to be a one-off, although it was hugely popular with photographers and enthusiasts alike. On the first day, 24 September, Fragonset's 33103 *Swordfish* is pictured at Oulton Broad North working the 13.29 additional from Norwich to Lowestoft. Fellow Fragonset loco 31452 *Minotaur* had been expected to work on this day but was found to be out of fuel, necessitating the cancellation of the first two special services. Peak 45112 *The Royal Army Ordinance Corps* was attached to the set of stock to run with 33103, but was prevented from working because of paperwork issues related to the swing bridges at Reedham and Somerleyton. Instead, FM Rail-liveried 47832 *Driver Tom Clarke OBE* accompanied the Crompton.

For many, the highlight of the ACoRP event was the use of Harry Needle Railroad Company Class 20s 20096 and 20905 on the Sunday, providing the first use of a Class 20 on a scheduled passenger service since the final summer Saturday service to Skegness in 1993. The immaculate pair of Type 1s are pictured in Great Yarmouth before working the 14.33 to Norwich on 25 September. Although not authentic for these particular locos, one Class 20, 20088, did wear unbranded Railfreight triple grey livery in BR service. Note the extra fuel tank on 20905 from its time with Hunslet-Barclay for use on weedkilling trains.

Above: Hauling the last special shuttle service of the ACoRP weekend, Harry Needle's 20096 leads 20905 towards Oulton Broad North past Leathes' Ham with the 16.27 Lowestoft–Norwich, with Anglia Railways-liveried 47714 on the rear. With every available window taken, the passengers are clearly savouring the sound of the English Electric Type 1s! By 2021, 20096 had joined the Locomotive Services Limited fleet with 20107, both still passed for use on the main line but repainted into BR green, while 20905 was still owned by HNRC and had been painted into GBRf colours with 20901 for a previous contract to deliver 'S Stock' underground trains from the Bombardier factory in Derby to Transport for London.

Right: For the 2006 Lowestoft Airshow specials, the Cotswold Rail fleet had been expanded and the previous Anglia Railways Mk.2 coaches had been replaced by 'One' Mk.3s. Ex-Virgin 47810 *Porterbrook* is seen approaching Oulton Broad North on 27 July on the rear of the 5G91 10.40 Lowestoft–Norwich, with 47818 providing the power. The attractive signal controlled the junction with the East Suffolk Line, seen diverging to the left.

Above: The second loco-hauled service of the morning on 27 July, bringing visitors to Lowestoft for the airshow, was 1G94, the 11.57 Norwich–Lowestoft. This is pictured coming round the curve at Oulton Marshes under Borrow Road bridge, on the approach to Oulton Broad North, with 47810 *Porterbrook* clearly applying some power. 47818 can just be seen on the rear. The unusual brick tower of Burgh St Peter church, in the shape of four cubes of decreasing size, can just be glimpsed on the horizon above the rear loco.

Left: Presenting a uniform corporate image for 'One' Railway, 47818 is seen working the first loco-hauled service for the 2007 Lowestoft Airshow, 1G92 09.50 Norwich–Lowestoft, through an overcast Oulton Broad North on 26 July. Cotswold Rail's 47828 can just be seen on the rear. A large rail-served fuel depot was once located on the left where the East Suffolk line to Ipswich is seen diverging in this picture.

Still no sun but plenty of 'clag'! Cotswold Rail's 47828 *Joe Strummer* powers through Oulton Broad North with the 5G91 10.40 Lowestoft–Norwich on 26 July 2007, taking the empty stock back to Norwich to collect the next trainload of visitors for the airshow. Note the Driving Van Trailer (DVT) behind the loco, this being used as the brake vehicle in the set and replacing the previous Anglia Railways DBSOs. The brick footbridge, linking Normanston Park with Lake Lothing, from where the previous picture was taken, can be seen beyond the rear loco.

In far better conditions than the previous day, 47828 *Joe Strummer* is pictured easing 5G91 10.40 Lowestoft–Norwich over Reedham swing bridge on 27 July 2007, creating a colourful scene with the 'One' Railway Mk.3 stock and DVT. 47818 can just be seen on the rear. The original single-track swing bridge was located directly to the left of the current bridge, with the new double track bridge constructed to the side in 1904. Network Rail undertook essential repair work on the bridge in October 2011, during a nine-day blockade. In 2021, Network Rail announced a consultation on work to replace the electrical and mechanical operating equipment of the bridge to ensure reliable future operation.

On Sunday 16 March 2008, Cotswold Rail 47818 and 47828 *Joe Strummer* were employed on a timing run to Lowestoft using Mk.3 coaches. Despite the less than ideal conditions, the Type 4s provided a welcome sight as an indication that regular loco haulage may return to the Wherry Lines, having been rather sporadic over the previous few years. 47818, now devoid of 'One' Railway stripes and branding, brings up the rear of the 5G34 13.48 Crown Point–Lowestoft and is seen passing Leathes' Ham between Oulton Broad North and Lowestoft.

On the return to Norwich, 47818 is pictured arriving into Oulton Broad North past the fine mechanical signalling, working the 5G35 15.22 Lowestoft–Norwich. The Mk.3s coaches had by this date received National Express branding as the franchise was rebranded from the previous 'One' Railway name.

Cotswold's 47828 *Joe Strummer* brings up the rear of the test service at Oulton Broad North, heading back to Norwich, past a solitary onlooker. The small housing development behind the platform was built on the site of Pope Brothers Garage; the road is named George Close after one of the brothers who ran the garage.

On 5 June 2008, 47818, now named *Emily*, worked a 5P33 12.15 Long Marston to Norwich Crown Point with six ex-Wessex Trains Mk.2 coaches, believed to be for potential use on the 'short set'. The ensemble is pictured in the yard at Norwich on the following day, with Wessex pink BSO 9525 nearest to the loco. Sadly, despite the timing runs and arrival of the stock, nothing materialised with the only reliable opportunity to experience loco haulage on the Wherry Lines at this time being the summer Saturday services to Great Yarmouth and the annual Lowestoft Airshow.

Much to the delight of enthusiasts and photographers, Advenza Freight-liveried 47375 was used on the 2008 airshow specials. With other Cotswold Class 47s being out of service or away from the area, the no-heat Brush Type 4 worked with regular 47818, leaving 47813 for 'Thunderbird' duties. This is 1G92, the 09.48 Norwich to Lowestoft, at Oulton Broad North on Thursday 24 July. 170206 can just be seen beyond the station returning as empty stock to Norwich with 156407 as the 5G61 10.20 Lowestoft to Norwich.

The evening service back to Norwich, 1G95 18.00 from Lowestoft, storms through Haddiscoe on 24 July with 47818 at the helm and Advenza 47375 at the rear. Note the cranes behind the rear coaches in the Environment Agency depot, used for work on the embankments of the River Waveney, seen behind the train, and the New Cut, branching off to the left to provide a shorter route by water towards Reedham and Norwich.

On Friday, 25 July, the first hauled service, 1G92 09.48 Norwich–Lowestoft, was again powered by 47375. The ensemble is seen easing through Reedham. Once named *Tinsley Traction Depot Quality Assured* while part of the Railfreight Distribution fleet, 47375 was exported to Hungary in 2015 for Continental Railway Solutions, where it operates charter services. The second hauled service on this day, the 1G94 11.55 Norwich–Lowestoft, was the last UK working for the popular Brush Type 4.

The fine array of semaphore signals at Oulton Broad North are shown to good effect as 47818 storms through with 1G95, the 18.00 Lowestoft to Norwich, the last loco-hauled service for the 2008 Lowestoft Airshow, on 25 July. 47375 can just be seen on the rear. The last remaining airworthy Vulcan, XH558, had featured in the air display during the afternoon. Introduced in 1960, the delta-winged nuclear bomber flew between 2007 and 2015, thanks to fundraising from the Vulcan to the Sky Trust.

By the summer of 2009, a major change to the traction seen on additional and loco-hauled services had taken place, with the contract passing from Cotswold Rail to DRS. While this brought the very smart DRS Class 47s to the area, the Cumbria-based company gradually acquired many of the Cotswold Rail Type 4s, bringing familiar locos to Norfolk under a new operator. The junction with the East Suffolk Line to Ipswich can be clearly seen in this illustration of 47501 *Craftsman* heading 1G92 09.48 Norwich–Lowestoft, the first Lowestoft Airshow loco-hauled additional of 2009, on 23 July. 47802 *Pride of Cumbria* was on the rear.

Later on 23 July 2009, 47501 *Craftsman* sweeps round the curve at Oulton Marshes towards Oulton Broad North station while working the 1G94 11.55 Norwich–Lowestoft airshow additional. 47802 *Pride of Cumbria* can just be seen on the rear, as can the excavators working on embankment and drainage work by the River Waveney. In 2016, 47501 was sold to Locomotive Services Limited for use on charters; based in Crewe, it was returned to original two-tone green livery.

Bracketed by semaphore signals, 47501 *Craftsman* takes the Lowestoft line at Reedham while working 1G92, the 09.48 Norwich–Lowestoft airshow additional, on the second day of the event, 24 July 2009. Reedham was a popular location for photographers for many years, with the appeal of the traditional signalling, fine signal box and telegraph poles creating an atmosphere of a country junction from the past.

A popular lineside location by the bridge next to Reedham signal box gradually became more overgrown over the years but still offered a good view in 2009. 47501 *Craftsman* is pictured easing towards the junction with the Berney Arms line while working the 1G94 11.55 Norwich–Lowestoft airshow additional, again on 24 July. 47802 *Pride of Cumbria* is again on the rear. The Railway Tavern, located adjacent to the station and just out of sight to the left on this picture, provided good business for services stopping at the Norfolk village for a number of years, particularly during their popular beer festivals. However, the pub was closed and converted into apartments in 2006.

Two contrasting modes of transport. The journey from Norwich to Lowestoft is quicker by rail than water, although the New Cut, a canal linking the rivers Waveney and Yare, which was completed in 1832, made the journey by boat quicker. However, the project was not a financial success with the New Cut sold to the developer Sir Samuel Morton Peto in 1842, who then opened the railway parallel to the waterway in 1847. It remained in railway ownership until nationalisation in 1948. 47501 *Craftsman* storms past a row of boats moored on the New Cut at Haddiscoe, while working 5G96 16.50 Norwich–Lowestoft on 24 July 2009, taking the empty stock to collect visitors heading home from the airshow.

Without doubt the rarest traction to work the 'short set' in its 25-year history was DRS Class 20, 20304, which was believed to be the first appearance of a Class 20 on a regular scheduled passenger train for 17 years. By 2010, the 'short set' had settled into a regular pattern operating most days. On Monday 8 February, 47712 and 47832 had been deployed, but news spread during the day that 20304 would replace 47832 on the Tuesday. As promised, the Type 1 operated nine round trips to Lowestoft and Great Yarmouth, paired with 47712 *Pride of Carlisle*. 20304 is pictured in Lowestoft coupled to DVT 82112, preparing to work 2J69, the 08.42 to Norwich.

DRS' 47712 *Pride of Carlisle* departs from Norwich with 2P10, the 09.36 to Great Yarmouth, with 20304 on the rear. The spire of Norwich Cathedral can be seen behind the exhaust fumes from the Class 47. Note the white horizontal pipes between the body and fuel tank, which were unique to the ex-ScotRail Class 47/7s. An example of the usual traction for this route, 156402 can be seen on the left; the Metro-Cammell built Super Sprinter wore advertising liveries for Anglia in Bloom and Chapelfield shopping centre while part of the Crown Point fleet.

Somewhat overlooked after the appearance of 20304, but still very unusual, DRS' 57004 was employed on 16–19 February 2010, and is pictured at the buffers in Great Yarmouth on 17 February after working the 2P30 18.40 from Norwich, with 47712 *Pride of Carlisle* on the rear. In subsequent years, the requirement for all locos operating the 'short set' to have Driver's Reminder Appliance (DRA) fitted along with remote fire extinguishers to enable the rear loco to be running while unmanned, prevented the appearance of more unusual traction.

The weather may not be ideal for photographs, but the semaphore signals still look impressive at Oulton Broad North Junction as 47832 *Solway Princess* storms through while working the first loco-hauled airshow additional of 2010, 1G92 09.48 Norwich–Lowestoft, on 12 August. 47501 is on the rear. The Type 4 has worn a wide range of liveries since privatisation, including Great Western green, FM Rail black, Victa Westlink, Stobart-branded DRS, DRS compass, Northern Belle and West Coast Railways maroon.

Making a contribution by helping to bring the many thousands of visitors to the 2010 Lowestoft Airshow, 47832 *Solway Princess* is pictured in Reedham on the second day, 13 August, while working the 1G94 11.55 Norwich–Lowestoft additional, with 47501 on the rear. Note the gap between the tracks in the station, which was necessary to accommodate an island platform; this was removed when the line was doubled at the turn of the 20th century.

On 21 June 2011, 47802 *Pride of Cumbria* waits to be started up at Great Yarmouth, having arrived on the rear of the 2P32 17.32 Norwich to Great Yarmouth, behind 47712 *Pride of Carlisle*. On the previous day, the '47s' and 'short set' had started operating morning and evening peak services to Great Yarmouth to cover for a damaged unit following a collision on the Sudbury branch. 47802 could always be identified within the DRS Class 47 fleet owing to the unusual oval buffers.

DRS-owned 47712 *Pride of Carlisle* is pictured departing from Norwich while working 2P12, the 08.35 Norwich–Great Yarmouth, on 27 June 2011, with 47802 *Pride of Cumbria* on the rear. 47712 was previously part of the fleet of ScotRail Type 4s deployed for push-pull operation between Edinburgh and Glasgow Queen Street, being named *Lady Diana Spencer* in 1981. It has now been preserved by the Crewe Diesel Preservation Group with a return to the attractive ScotRail livery, and in 2021 was active on the main line under Locomotive Services Limited.

Like many of its classmates, 47841 was once part of the large fleet of Class 47/8s fitted with long range fuel tanks and deployed on CrossCountry services and, as such, could be found operating across the network. After Virgin Trains replaced these loco-hauled services with Bombardier-built Voyagers, 47841 passed between various operators including Freightliner and Harry Needle, before joining the DRS fleet in 2008. Despite its reputation for not being the most reliable loco in the fleet, it still looked smart pictured in Norwich on a dull 30 June 2011, before working the 17.32 to Great Yarmouth. The loco is now owned by Jeremy Hosking's company, Locomotive Services Limited, being on static display at the One:One Collection in Hornby's former factory in Margate.

The power is applied on 47810 *Peter Bath MBE 1927-2006* at Norwich on 5 July 2011, as 2P12 08.35 to Great Yarmouth departs with 47802 *Pride of Cumbria* on the rear. This image shows how the DRS Class 47s were always very clean and presentable. Peter Bath was founder of the Bournemouth-based travel company, Bath Travel, with the name applied in April 2011. The name *Captain Sensible* had been worn by the Class 47 under its previous operator, Cotswold Rail.

The Lowestoft Airshow specials did not get off to a good start in 2011, with the first loco-hauled service being covered by a Class 153 DMU as the Class 47s were still being fuelled on Crown Point depot in Norwich. However, the second run, 1G94 11.27 Norwich–Lowestoft, caught the last of the sunshine on the Thursday morning, 11 August, with 47802 *Pride of Cumbria* at the helm and 47810 *Peter Bath MBE 1927-2006* on the rear, pictured powering through Oulton Broad North.

In a very welcome moment of sunlight, 47810 *Peter Bath MBE 1927-2006* storms through Haddiscoe with the 1G93 15.40 Lowestoft–Norwich, taking the first load of visitors home after a day at the seaside. The supports for the swing bridge which once carried the Yarmouth Southtown to Beccles line over the River Waveney can just be seen behind the penultimate coach. This line closed in November 1959, reportedly because of the maintenance costs of this swing bridge and the one at Beccles, with traffic diverted via Lowestoft and the now closed Lowestoft to Yarmouth Southtown line.

The evening sun illuminates 47818 and 47810 in Great Yarmouth on 14 September 2011 before they return with 2P33, the 18.17 to Norwich. On this day, the duo had worked 2J74 12.06 Norwich to Lowestoft and return, and 2P24 14.36 to Yarmouth, in addition to the regular morning and evening peak services. Another loco which was once part of the large CrossCountry fleet, 47818 was for a while clearly identifiable as the only member of the class to wear 'One' livery but, like many of the former Cotswold Rail Type 4s, it had by this time passed to DRS ownership.

Showing the range of Class 47s which could be found on the 'short set' at the time, 47501 *Craftsman* is pictured with 47810 *Peter Bath MBE 1927-2006*, with the same working, 2P33 18.17 to Norwich, on 21 September 2011. Four return trips to Yarmouth and three returns to Lowestoft were completed by the pair on this day. In addition to these services, the DRS Class 47s could also be found on through workings from London Liverpool Street to Yarmouth on summer Saturdays until 2014, known as the 'drags' as the Class 47s dragged full rakes of coaches through to the coast.

Despite the large number of visitors to the annual Lowestoft airshows, the organisers struggled to break even, with visitor donations never covering costs. In the final year, the event was moved from a Thursday and Friday in August to a weekend in June. The sugar beet factory at Cantley, which can be seen for many miles across the flat Norfolk marshes, is visible from the vantage point of the A143 bridge at Haddiscoe, with the first airshow special of the day on 23 June 2012, 1G92 09/28 Norwich to Lowestoft, storming along the New Cut. 47828 is providing the power, with 37419 at the rear. In February of this year, the East Anglian franchise had passed from National Express to Abellio, a subsidiary of the Dutch state operator, and was branded as Greater Anglia. The new company logo can just be seen on the white stripes of the Mk.3 coaches.

The final Lowestoft Airshow in 2012 provided the opportunity to experience two of DRS' newly repainted and very smart Class 37/4s on the additional services, along with the more familiar Class 47s, on two sets of stock unlike the previous single set. The prospect of a Class 37 crossing the Reedham swing bridge in the sun clearly could not be missed and produced an impressive gallery of around 30 photographers on the morning of Saturday 23 June 2012, many armed with step ladders to get the perfect view. Immaculate 37419 makes an impressive sight and sound as it powers over the bridge while working the 5G91 10.28 Lowestoft to Norwich.

The second hauled set for the Lowestoft Airshow produced notoriously unreliable 47841, which failed the following day causing the cancellation of the first Sunday service. The former Virgin 'Brush 4' is still going strong while working the 1G64 10.25 Norwich to Lowestoft at Reedham on 23 June 2012 and approaching the swing bridge. 37425 was on the rear.

Looking like a Yarmouth summer Saturday service, this is in fact the 1G94 11.25 Norwich to Lowestoft, the third and final outwards service for the Lowestoft Airshow on Saturday 23 June 2012, seen in the classic location at Reedham. 37419 is on the rear. Cattle pens were once located just beyond the signal on the left; the top of a loading gauge can just be seen to the left of the signal – another relic from a past era on the railway.

After a week of much speculation about the traction for the Lowestoft Airshow specials of 2012, especially in light of the fact that two sets of stock were in use, the Saturday morning provided final confirmation that 37419 and 37425, newly overhauled and added to the DRS fleet, were to visit the seaside. The event provided the first Class 37-hauled service train to Lowestoft for some 18 years. 1G97 17.45 Lowestoft to Norwich was delayed after an incident on the train which required the police to attend; this caused the service to be held outside Oulton Broad for some time. 37425 tries to make up time at Haddiscoe, storming back to Norwich.

The 'short set' between duties number one: when not in use, the locos and stock could often be found in the 'low level' yard in Norwich. However, on Monday 22 April 2013, 47805 and 47828 (out of sight at the rear of the coaches), which had hauled the set on the previous Friday, could be found in the company of 47810 *Peter Bath MBE 1927-2006*, 47802 *Pride of Cumbria* and 47818! Although the DRS Class 47s were a regular sight at the time, seeing five together in Norwich was definitely a rarity. 47805 and 47828 were not used again in service until the following Friday, 26 April.

The 'short set' between duties number two: on 9 July 2013, 47818 and 47813 were found in the company of BR Green 31190 in Norwich. The Class 47s were used for three Yarmouth and two Lowestoft returns on the following day. 31190 had returned 73210, 50026, 47596 and 47579 from the East Lancashire Railway to the Mid Norfolk Railway on Monday 8 July, after featuring at a diesel gala on the former line. The Class 31 worked light engine as the 0Z46 10.15 to Crewe Heritage Centre, later on the morning of the photograph.

Above: On 18 July 2013, 47813 could be found enjoying the seaside sun at Great Yarmouth before working 2P33 18.17 to Norwich, this being the last of five trips to the Norfolk and Suffolk coast during the day. 47853 was on the rear. Note the silver Class 57-style roof grilles on the veteran Brush Type 4. 47813 was still operational on the main line in 2021, initially as part of the Rail Operations Group fleet, but then with West Coast Railways, following its sale.

Right: While the 'short set' was being used on a more frequent basis by the summer of 2013, with operation throughout the day, the Class 47s only saw action if services could not be operated by a DMU. The traditional railway infrastructure is clearly in evidence in this shot of 2J80, the 14.58 Norwich–Lowestoft, on 6 August nearing its destination with 47841 leading 47818. At this time, the fine semaphore signals were still controlled by Lowestoft signal box, seen on the right. They were finally replaced with modern colour light signalling in February 2020. The signal box was opened to the public as part of the Heritage Open Days in September 2019, providing a fascinating opportunity to see the traditional signalling still in action. The Lowestoft Community Rail team won the national Community Rail Award for Tourism and Heritage for these signal box tours.

Left: A very traditional railway scene on the modern privatised railway as a locomotive-hauled service traverses complex pointwork and passes mechanical shunt signals and older-style speed restriction signs. 47841 and 47818 arrive at Lowestoft on 6 August 2013, with the 14.58 from Norwich. The Iron Bridge can be seen in the background, which was once a popular photographic vantage point. However, by this time, it had been closed to the public, which was a sad but very understandable decision, given the condition of the bridge.

Below: 47818 was a regular on the 'short set' initially under Cotswold Rail ownership, and later with DRS. It is pictured on the rear of 2J80, the 14.58 Norwich–Lowestoft, on 6 August 2013, about to arrive at Britain's most easterly station. 47841 is on the front. The imposing brick building in the background was once Tuttles department store; part of the attractive building on Station Square is now occupied by The Joseph Conrad, a Wetherspoons pub.

On Friday 30 August 2013, the availability of Crown Point DMUs was clearly low enough to necessitate the use of the 'short set', this being the first outing for the Type 4s that week. 47841 is pictured with 47828 in Great Yarmouth with the regular working, 2P33 18.17 to Norwich. 47841 had previously worn the names *The Institution of Mechanical Engineers* and *Spirit of Chester*, but, by this time, was un-named.

47828 is clearly providing the power as 47841 tails 2P22, the 13.36 Norwich–Great Yarmouth, past Whitlingham Junction on 31 August 2013. Unusually for this time, the 'short set' was working on a Saturday to cover for unavailable units, necessitating the cancellation of the booked Class 47-hauled Yarmouth to London 'drags'. Five trips to Yarmouth and one return to Lowestoft were completed by the set on this day. The photograph was taken from the former Whitlingham station footbridge. The station closed in September 1955, but the footbridge still provides access to Thorpe Marshes, a popular destination for dog walkers. The platforms of the former station extended either side of the rear loco, as far as the points for the Sheringham line.

By the summer of 2014, the Greater Anglia Mk.3 coaches and DVT used in the 'short set' had been replaced by Mk.2 coaches hired from DRS in the form of two TSOs and one BSO. The matching blue stock can be seen here in Norwich on a dull 11 July after the arrival of 2P21, the 13.17 Yarmouth–Norwich, with 47805 *John Scott 12.5.45-22.5.12* and 47501 *Craftsman* providing the traction. Note the points under the second coach which were once used for run-round manoeuvres on platform 4.

Although still not used every day, by 2014, when the 'short set' was in use, three Lowestoft and four Great Yarmouth trips were completed. This diagram continued until the operation finished in September 2019, although at this time, it usually only ran on Mondays to Fridays, with Saturday operation added later. 47853 *Rail Express* can be clearly spotted owing to the gold buffers, being pictured in Great Yarmouth on 22 July with 47501 *Craftsman*, preparing to work the 2P33 18.17 departure to Norwich.

Other than the latest liveries, a scene which has not changed for the last 30 years: a Class 47/4 with Mk.2 stock passing semaphore signals. 47501 *Craftsman* leads 47853 *Rail Express* with the 'short set', deputising for the usual London to Great Yarmouth 'portions' because of engineering work in Ipswich, with 1V18 12.03 Norwich–Great Yarmouth at Cantley on 9 August 2014. The Down Home No. 21 co-acting signal, seen behind the Class 47, was very rare, featuring two arms, with the higher one simply repeating the lower (out of sight behind the loco) to aid visibility. Like all other signals on the line, it was replaced as part of the Wherry Line resignalling project.

The perfect rural setting: 47853 *Rail Express* ambles off the Berney Arms line at Reedham with 1V43 13.10 Great Yarmouth to Norwich with 47501 *Craftsman* bringing up the rear on 9 August 2014. The tracks to the right curve round to the swing bridge and then on to Lowestoft. This service would have usually continued to Liverpool Street with a full ten-coach set but, owing to engineering works at Ipswich, the usual 'portions' were worked by the 'short set', providing a rare appearance of top-and-tailed Class 47s and Mk.2s on the branches on a Saturday.

A real gem of a location, with more period charm than many preserved railways. There were few locations on the main line which still boasted semaphore signals and telegraph poles by 2014. 47501 *Craftsman* brings up the rear of the 1V43 13.10 Great Yarmouth to Norwich, with 47853 *Rail Express* providing the traction, as the ensemble joins the Lowestoft line at Reedham Junction on 9 August. The centre of Reedham village is situated south of the junction (to the left of the photograph) with an attractive riverfront beyond the main street.

Well worth the early alarm call to get the shot on 29 August 2014, as 47501 *Craftsman* leads 47813 *Solent* over Reedham swing bridge while working 2J67, the 07.47 Lowestoft–Norwich. This was the first passenger working of the day, with the set going to Lowestoft empty as 5J67 06.40 Crown Point–Lowestoft. The land on the far side of the bridge is relatively inaccessible, with the only way of crossing the River Yare by car or foot being the Reedham Ferry, located a mile inland from the swing bridge.

Running late after brake problems at Brundall Gardens, the next service on this day missed the path on the single line to Acle. 47813 *Solent* is pictured storming away from Brundall working 2P12 08.36 Norwich–Great Yarmouth at Cuckoo Lane with 47501 *Craftsman* on the rear on 29 August 2014. The junction with the line to Reedham and Lowestoft can just be seen in the distance. Two other Class 47s were in action in Norfolk on this day, with 47367 and 47596 taking part in the Mid Norfolk Railway summer diesel gala.

Catching up after delays to the previous working on 29 August 2014, 47501 *Craftsman* storms through Stracey Arms working 2P13, the 09.17 Great Yarmouth–Norwich, with 47813 *Solent* on the rear. The location is named after Stracey Arms Windpump, located just off the picture to the left; the windpump was built in 1883 for Sir Henry Stracey. The busy A47 'Acle straight' can just be seen on the left behind the telegraph poles, with the road running parallel to the railway across the marshes between Yarmouth and Acle.

Clearly showing the need for the 'short set', crowds of holidaymakers join 2P21 at Great Yarmouth on 29 August 2014, the 13.17 to Norwich, with the brake van providing excellent storage for a number of bicycles. Although the days of direct summer Saturday loco-hauled services from the Midlands and the North have long gone, trains to Yarmouth were still very busy on Mondays and Fridays, with more weekend trips to the Norfolk resort as opposed to the traditional Saturday to Saturday week by the coast. 47501 *Craftsman* will provide the power, with 47813 *Solent* on the rear.

The bright morning sun on 27 March 2015 highlights the attractive Northern Belle livery of 47790 *Galloway Princess* while heading 2P12, the 08.36 Norwich–Great Yarmouth service, at Cuckoo Lane, just after leaving the Lowestoft line at Brundall. Note the cast crest on the cabside of the loco. 47818 brings up the rear. Before being purchased by DRS, 47790 had been part of the large Crewe-based fleet of Rail Express Systems Class 47/7s usually deployed on postal workings. It moved to Motherwell in 1998 for use on Scottish sleepers before being repainted into EWS livery in 2003.

Later on the same day, 27 March 2015, Northern Belle-livered 47790 *Galloway Princess* works the 2P20 12.36 Norwich–Great Yarmouth at Brundall with 47818 on the rear. 47790 was one of two Class 47s repainted when DRS took over the contract to provide traction for the Northern Belle luxury train, 47832 *Solway Princess* being the other. Seeing one of the Northern Belle locos hauling something other than the matching excursion stock was relatively unusual.

The end of the use of Class 47s on the 'short set' resulted in the final workings of two Type 4s. The last run for 47805 was 2J88 19.00 Norwich–Lowestoft on Friday 12 June, with East Anglian regular 47818 bowing out on the same working on Monday 15 June. On the Friday, 47813 is pictured in Great Yarmouth before departing with the 2P33 18.17 Yarmouth–Norwich and about to leave the Norfolk coast for the last time. 47805 was out of sight on the rear.

After months of rumours and speculation, it finally happened! For the first time in just under 20 years, a Class 37 hauled a regular passenger service to Lowestoft and Great Yarmouth. The last day of Class 47 operation was Tuesday 16 June 2015, with the final working hauled by 47813 on 2J67 07.47 Lowestoft to Norwich. The rest of the diagram was covered by DMUs, and then on Wednesday 17 June, 37425 *Sir Robert McAlpine/Concrete Bob* worked the 2J67 07.47 Lowestoft to Norwich. 37405 and 37425 are pictured in Norwich later in the day preparing to work 2P32, the 17.36 to Great Yarmouth.

And then the sun came out! On 27 June 2015, 37405 storms towards Whitlingham Junction while working 2C54, the 10.25 Norwich–Great Yarmouth, with 37425 *Sir Robert McAlpine/Concrete Bob* bringing up the rear. This was one of four returns to the Norfolk coast operated by the popular Type 3s on this day, with the 'short set' now covering these non-stop additional Saturday services during the summer. Girlings Lane crossing can just be seen behind the rear loco; on this road there is a plaque commemorating the Thorpe railway disaster of 10 September 1874, when 25 people were killed by a head-on collision near this spot.

37405 leads partner 37425 slowly into Brundall while working 2C58, the 12.18 Norwich–Great Yarmouth, which was another non-stop summer Saturday service on 27 June 2015. This was the second Saturday since Class 37s replaced Class 47s on the 'short set'. When the original line to Yarmouth via Reedham opened in 1844, Brundall station was built on a road which went from the village to a ferry over the River Yare to Coldham Hall in Surlingham. Although the ferry stopped operating in the 1970s, the road is still busy providing access to many boat yards.

37425 *Sir Robert McAlpine/Concrete Bob* storms away from Brundall on 30 June 2015 while working 2P33, the 18.17 Great Yarmouth–Norwich, with some help from 37405 on the rear. This pair of locos worked every day from the start of Class 37 operation on 17 June until 18 July. 37425 became known locally as 'Angry Bob' to acknowledge its rather impressive engine sounds, these regularly shattering the peace in villages along the line, particularly after stops at the intermediate stations such as this.

On 17 July, 37425 *Sir Robert McAlpine/Concrete Bob* brings up the rear of 2P32, the 17.36 Norwich to Great Yarmouth, with 37405 on the front. The Mk.2 coaching stock featured through wiring for Blue Star multiple working, enabling both locos to be controlled from the front. Remote fire extinguishers were fitted to enable the rear loco to be powered while unmanned. The combined 3,500hp provided more than enough power to keep to time on the start-stop diagram!

Do the commuters on 2J67, the 07.47 Lowestoft–Norwich, appreciate the quality traction taking them into the city to the office? 37419 *Carl Haviland 1954-2012* rumbles over Reedham swing bridge with some assistance from 37425 *Sir Robert McAlpine/Concrete Bob* on the rear on 3 August 2015. Both locos had worked to Lowestoft during the 2012 airshow but, at the time, it was not known that they would later become regulars on the Wherry Lines. Although these routes have now been resignalled, Reedham swing bridge signal box, along with the box at Somerleyton, are still in use as they control the operation of the bridges, which regularly open for the passage of river traffic.

37425 *Sir Robert McAlpine/Concrete Bob* storms away from Brundall with some help from 37419 *Carl Haviland 1954-2012*, shattering the peace at Cuckoo Lane, while working 2P12, the 08.36 Norwich–Great Yarmouth, on 3 August 2015. 37425 had been painted into BR large logo blue livery in 2005 by its previous operator EWS and named *Pride of the Valleys* to commemorate the end of loco-hauled services between Cardiff and Rhymney, while in July 2009, 37419 became one of only two Class 37s to receive the DB Schenker red livery.

The length of the passing loop at Acle, which was extended in 1960 to accommodate loco-hauled summer Saturday services, can be seen as 37425 *Sir Robert McAlpine/Concrete Bob* leads 37419 *Carl Haviland 1954-2012* while working 2P18, the 10.36 Norwich–Great Yarmouth, later on the same day, 3 August 2015. The vans on the right are on the site of the former goods yard which closed in 1969.

37419 *Carl Haviland 1954-2012* leads 37425 *Sir Robert McAlpine/Concrete Bob* while working 2P17, the 11.17 Great Yarmouth–Norwich, towards Brundall where the single Acle line joins the double track from Reedham on 3 August 2015. Until the DRS Class 37s replaced the Class 47s earlier in 2015, ETH-fitted Class 37/4s were rarely seen in East Anglia, with passenger appearances confined to charters only.

37425 *Sir Robert McAlpine/Concrete Bob* slows for the stop at Brundall while working 2P20, the 12.36 Norwich–Great Yarmouth, with 37419 *Carl Haviland 1954-2012* bringing up the rear, later on 3 August. Note the staggered platforms, with the Norwich-bound one being the other side of the footbridge behind the photographer; this was necessary as the station was built on a narrow plot of land.

The sun catches the nose of 37425 *Sir Robert McAlpine/Concrete Bob* while storming through Buckenham on 22 August 2015, with a Saturday non-stop service, the 2C54 10.25 Norwich–Great Yarmouth, with plain blue 37422 on the rear. Situated between Brundall and Cantley, Buckenham is a peaceful rural location with services only stopping at weekends, although these can be useful for visiting Strumpshaw Fen nature reserve, which is a one-mile walk from the station along quiet country lanes. A ferry across the River Yare operated from Buckenham until the early 1940s; the presence of the ferry may have justified the building of the station in a very sparsely populated area.

The work of the station adopters can be clearly seen on a number of Wherry Line stations, in particular Cantley, which boasts attractive and well-maintained floral displays. However, did the passengers storming through on the 2C55 11.14 Great Yarmouth–Norwich on 22 August 2015 appreciate the flower-filled boat? 37422 and 37425 are providing the traction. Like its neighbour Buckenham, Cantley was also a very small hamlet when the line was opened in 1844, but has since grown, owing to the building of the sugar beet factory.

The telegraph poles can still be seen in the background as 37425 powers through Reedham Junction later on 22 August 2015, working the 2C58 12.18 Norwich–Great Yarmouth, with some help from 37422 on the rear. This scene has radically changed with the removal of the semaphore signals, simplification of the track layout and the replacement of the 'short set' with new Stadler bi-mode units. The passengers on this service are clearly savouring the experience of English Electric traction in the knowledge that it would not last for ever!

A moment of fame for 37422, as the unbranded DRS '37/4' slows to join the Lowestoft line at Reedham Junction while working the 2C59 12.55 Great Yarmouth–Norwich, as it was being observed by a gallery of 12 photographers on 22 August 2015. The Type 3 had joined the Anglia fleet on the previous Saturday, operating its first passenger service for seven years and the first under DRS ownership. It did not receive the usual DRS branding as the bodywork was felt not to be of an adequate standard on which to apply the vinyls seen on other locos. St John the Baptist Church can be seen in the top left, which is located to the north east of Reedham village. The church is constructed using a large amount of reused Roman materials but was gutted by a fire in 1981.

Later on 22 August, unbranded 37422 eases through Reedham and past a fine array of mechanical signalling while working 2C63, the 14.55 Great Yarmouth–Norwich, with some help from 37425 *Sir Robert McAlpine/Concrete Bob* on the rear. After resignalling, Reedham Junction signal box, seen on the right, was generously gifted by Network Rail to the North Norfolk Railway for use as a signalling simulator. By 2021, the heritage railway had almost raised adequate funds to start the process of moving the structure to Holt.

Coaching stock changes number one: with the regular DRS Mk.2 coaches scheduled to be taken out of service during the following week for maintenance, 37405 is pictured in the 'low level' yard in Norwich on Friday 22 April 2016 with a single coach, TSO 5919. On this day, 37419 *Carl Haviland 1954-2012* and 37422 were operating the usual 'short set' diagram.

Coaching stock changes number two: to cover for the Mk.2 stock being taken out of service for maintenance, the 'short set' was temporarily reformed with Abellio Greater Anglia Mk.3 stock, including DVT 82152. The set is seen in the 'low level' at Norwich on Sunday 24 April ready for another busy week of services to Great Yarmouth and Lowestoft, with traction from 37419 and 37405.

Have the customers at The Rushcutters Arms in Norwich chosen to sit outside to enjoy the passing English Electric entertainment? 37419 *Carl Haviland 1954-2012* leads 37422 (out of sight at the rear) as they approach their destination with 2P33, the 18.17 Great Yarmouth–Norwich. The Grade II listed pub was built in the 16th century and was previously called the Boat and Bottle. It is now run by a national chain, but is still a popular destination for food and drink, especially in the summer months, with plenty of outside space overlooking the River Yare.

Following the collision of Turbostar 170204 with a tractor near Thetford on 10 April 2016, the damaged unit was covered by a second 'short set' hauled by DRS Class 68s as no suitable diesel units were available to hire, with services starting on Monday 4 July. The cost of the operation was covered by insurance, this including the hire of the rolling stock, two examples of DRS' newest class of locos, plus drivers from DRS as Greater Anglia staff did not sign this type of traction – not an insignificant expense! 68016 *Fearless* arrives in Norwich while working 2P29, the 17.17 from Great Yarmouth, on 27 July. 68019 *Brutus* is almost out of sight on the rear. Note the Anglia Railways Mk.2 stock hired from Riviera Trains, although the leading BSO never worked with Anglia Railways, and had been painted to match the other vehicles.

Completing its second round trip to the coast on 9 August 2016, 68016 *Fearless* leads 68019 *Brutus* through Whitlingham Junction with the late running 2P11 08.45 Great Yarmouth–Norwich, with the line to Sheringham visible behind the first coach. The Class 68s were built by Stadler Rail (previously Vossloh) in Spain, being a UK derivative of the Stadler Eurolight design. The locos have become very popular with enthusiasts, possibly because of their impressive sounding Caterpillar engines, giving them the nickname of 'Cats'.

Back on schedule after the late running inwards working from Great Yarmouth, 68019 *Brutus* leads 68016 *Fearless* through Whitlingham Junction, working 2J70, the 10.05 Norwich–Lowestoft, on 9 August 2016. 68019 is the first in the fleet of Class 68s which are now deployed on TransPennine Express services between Liverpool Lime Street and Scarborough, hauling CAF-built Mk.5 coaches. These are the first newly built passenger coaches since the BR Mk.4s from the early 1990s.

Right: Later on 9 August 2016, the original 'short set' is seen at Brundall with regular performers 37419 *Carl Haviland 1954-2012* and 37422 working 2P20, the 12.36 Norwich–Great Yarmouth. 37419 was built as D6991 with introduction to service in June 1965 at Cardiff Canton. It was converted to a Class 37/4 with the addition of electric train heat in 1985, before moving to Scotland to haul passenger services in the highlands. A transfer to Tinsley depot in Sheffield took place in 1991, with regular use on passenger services around Manchester.

Below: The old and the new: the semaphore signals and manual signal box at Lowestoft had a limited life at this time, on 10 August 2016, but hopefully 68019 *Brutus* will be on the rail network for a few years to come. The Caterpillar-powered loco is seen working 2J73, the 10.57 to Norwich, with 68016 *Fearless* on the front. Note that the Iron Bridge, which once crossed the yard in the distance, had by this time been removed, with the yard itself being heavily overgrown after the withdrawal of the short-lived Enterprise freight service to Aberdeen.

68016 *Fearless* is pictured bringing up the rear of 2J80, the 14.55 Norwich–Lowestoft, on 10 August 2016, departing Oulton Broad North, with 68019 *Brutus* on the front. The pair of Class 68s were covering for the usual Class 37s after an incident at Brundall earlier in the day. Edging stones for the new platform can be seen on the right, part of a project to move the Norwich-bound signal further from the notoriously busy level crossing to allow trains to arrive at the station before the crossing gates were lowered.

On 12 August, 68019 *Brutus* and 68016 *Fearless* could still be found deployed on the second 'short set' diagram with 68019 powering the ensemble towards Brundall Gardens with 2J70, the 10.05 Norwich–Lowestoft. Access to this location is via a footpath from Brundall Gardens station; on the opposite side of the track are some well-maintained allotments (just out of sight to the right), and Brundall Countryside Park.

68016 *Fearless* is pictured storming through Cantley later on 12 August 2016, with 68019 *Brutus* on the rear, working 2J83 15.48 Lowestoft–Norwich, past the magnificent floral displays maintained by the active station adopter team. The boat on the right is named *The Rose of Cantley*. The 14.55 Norwich–Lowestoft and return were usually worked by the Class 37s but 37419 and 37422 were stopped after working the 11.17 from Great Yarmouth on this day, necessitating the extra trip for the Class 68s.

Also on 12 August, 68016 *Fearless* powers away from Great Yarmouth working the 2P29 17.17 departure back to Norwich and past the fine signal box. Following resignalling of the Wherry Lines, some disused signal boxes have been either listed or earmarked for further use elsewhere, but being of brick construction, the 'Yarmouth Vaux' box cannot be easily moved. The resignalling project has also resulted in platform 1, seen on the left, being removed, leaving Great Yarmouth with three operational platforms.

The sharp modern lines of the DRS Class 68s present a powerful image as 68022 *Resolution* leads 68024 *Centaur* into Norwich with 2P29, the 17.17 from Great Yarmouth, on 19 September 2016. After the initial dominance of 68016 and 68019, the locos used for these services started to vary as maintenance was required, with a very impressive 17 different engines being used in the 14 months of operation. Each time a new loco appeared on the set, enthusiasts would arm themselves with a Wherry Lines Rover ticket and their red pen to mark off their first ride behind each of these impressive machines.

After working the 2C35 18.47 from Great Yarmouth, 68022 *Resolution* waits at Norwich on 19 September 2016, with 68024 *Centaur* at the other end. 68022 had replaced 68004 *Rapid* on the second 'short set', working its first services on this day. The front of a 12-car Class 321 set can just be seen on platform 1 on the right; these units were booked to visit Norwich on the 17.02 stopping service from Liverpool Street and 19.30 return, providing variation from the usual Class 90s and Mk.3 coaches used on other Great Eastern Main Line services at this time, but possibly not the same level of comfort!

68005 *Defiant* is pictured at Norwich on the rear of 2P34, the 18.06 to Great Yarmouth, on its first day working with Greater Anglia, 7 November 2016. 68002 *Intrepid* is on the front. Note the blue/grey stock which had replaced the previous Anglia Railways-liveried vehicles. Sadly, this set of authentically liveried stock never appeared on the other 'short set' with the BR large logo Class 37s, missing an opportunity to recreate scenes from the 1980s.

On the cold and damp evening of 12 December 2016, 68005 *Defiant* briefly pauses at Great Yarmouth before working 2C35, the 18.47 to Norwich via Reedham. 68025 *Superb* is on the rear. After the end of the contract with Greater Anglia, the Class 68s not hired to TransPennine Express could be found deployed on a range of duties, including intermodal freights, nuclear flask workings, engineering trains, Network Rail test trains and charters.

The one everyone was waiting for: after a major overhaul from virtually derelict condition, having been purchased un-restored from the Churnet Valley Railway, large logo 37424 *Avro Vulcan XH558* is pictured on its first full day in service with Greater Anglia. This was Monday 19 December 2016, with the loco seen at Norwich before storming out of the station with 2J88, the 19.02 to Lowestoft, with 37405 on the rear. 37424's first passenger service for over 16 years had been the 19.02 to Lowestoft on the previous Friday, 16 December. The impressive Type 3 is named after the Vulcan aircraft which visited the Lowestoft Airshow on several occasions.

Clearly identifiable by the WIPAC light clusters, 37423 *Spirit of the Lakes* pauses at dusk in Great Yarmouth before working 2P33, the 18.17 to Norwich, on 6 March 2017. 37425 *Sir Robert McAlpine/Concrete Bob* had worked the 17.36 from Norwich. As D6996, the Type 3 was new to Cardiff Canton in July 1965, and like the other Class 37/4s had electric train heat installed in 1985, moving to Glasgow Eastfield for use on passenger services in Scotland. Following withdrawal by EWS in 1999, it passed to Ian Riley Engineering and then West Coast Railways but was returned to service by DRS in 2007.

Seeing regular loco-hauled passenger services on a branch line in 2017 was quite something, but having a choice of traction for different services was unprecedented so many years after most other regular loco-hauled services had ended. At certain times, both 'short sets' could be seen together, as illustrated here on 28 April, with 37423 *Spirit of the Lakes* and 37403 *Isle of Mull* on the left with the 2J80 14.55 to Lowestoft, while 68018 *Vigilant* and 68003 *Astute* prepare to depart with the 2J78 14.05 to Lowestoft on the right, a service only covered by the Class 68s if a unit was not available. Note the Anglia Railways-liveried stock on the latter set which had returned by this date, replacing the blue/grey vehicles.

The Scottish Railway Preservation Society restored two of its Class 37s, 37025 *Inverness TMD* and ETH-fitted 37403 *Isle of Mull,* to main line standards with the intention of utilising them for charter work in Scotland. However, opportunities arose for the hire of both locos to freight companies, with 37025 working for Colas and DRS using 37403. The latter loco is pictured in Norwich on 28 April 2017, in the company of 37423 *Spirit of the Lakes*, after working 2P21, the 13.17 from Great Yarmouth. Note the West Highland terrier logo on the bodyside, which was originally applied to locos allocated to Glasgow Eastfield depot.

Above: Thanks to the two 'short sets' on Greater Anglia during 2016–17, day trips to locations such as Somerleyton, on the Lowestoft line, were possible with the outward journey featuring one class of locomotive and the return on another. On 3 May 2017, the outward trip featured 68018 *Vigilant* with 68003 *Astute* on the rear. The station building can just be seen on the right, it being grander than others on the line. This is explained as Somerleyton Hall was once home to Sir Samuel Morton Peto, who instigated the line from Reedham to Lowestoft. The author's wife and seven-month-old son can be seen on the platform, having just enjoyed their ride on the Class 68s!

Left: The return journey from Somerleyton on 3 May 2017. Other than the DRS coach, this could be a scene from the 1980s with a large logo Class 37 and semaphore signals. 37403 *Isle of Mull* rounds the curve approaching Reedham Junction while working 2J83, the 15.48 Lowestoft–Norwich. 37423 *Spirit of the Lakes* was on the rear. The signal is on the line from Great Yarmouth via Berney Arms which joins at the junction.

Above: 37419 *Carl Haviland 1954-2012* trails 37405 with 2P32, the 17.36 to Great Yarmouth, departing Norwich on 3 June 2017. This service was usually operated by a unit on a Saturday. Note the dent on the nose of 37419, sustained after striking a fallen tree on the Great Eastern Main Line during Storm Doris in February. The formation of the set includes DBSO 9705; these driving brake vans were used to control the Class 37s from the rear of the set, while operating for Northern on the Cumbrian Coast, but were never used in this way by Greater Anglia. It was believed that the East Anglian operator preferred the assurance of two locos on the set in case of a failure.

Right: On 9 June 2017, the powerful headlights on 68022 *Resolution* can be seen in Brundall as it leads 68024 *Centaur* working 2P28, the 16.38 Norwich–Great Yarmouth. The line between Norwich and Yarmouth via Reedham was the first in Norfolk to open, with services starting on the initial single track on 1 May 1844, this being doubled at the turn of the century. The more direct route via Acle, over which this service would run, was opened in 1883, cutting the journey by just over two miles.

Back in Norwich after the previous trip to the coast, 68024 *Centaur* trails 68022 *Resolution* with 2P34, the 18.04 to Great Yarmouth, on 9 June. The complex catenary which can be seen above the set was installed at the same time as rationalisation and resignalling, with the first electric services arriving into the city in 1987, initially hauled by Class 86s and later replaced by Class 90s. Crown Point depot, where most local rolling stock is maintained, is located just beyond the bridge in front of the leading loco.

37419 *Carl Haviland 1954-2012* brings up the rear of 2P32, the 17.36 to Great Yarmouth, this time with 37422 at the front on 14 June. A Class 170 Turbostar can be seen disappearing on the right on a service to Cambridge; these units were introduced in 1999 by Anglia Railways, but had all moved to Transport for Wales by February 2020 upon the introduction of the new Stadler units.

68024 *Centaur* leads un-named 68027 into Norwich with the 2P29 17.17 from Great Yarmouth, also on 14 June. 68027 has since been named *Splendid*, with both locos now being part of the TransPennine fleet. The large numbers above the signals indicate which platform they apply to, a handy reminder for drivers and photographers observing movements in and out of the station.

An Acle encounter as brand new 68027 waits for 37419 *Carl Haviland 1954-2012* to clear the single line to Great Yarmouth on 14 June while working 2P34, the 18.04 departure from Norwich. In the past, this line and the route via Berney Arms would have been pushed to breaking point on summer Saturdays with many additional holidaymaker trains. However, the sight of two loco-hauled passenger trains passing in Acle had not been seen for many years.

68024 *Centaur* makes an impressive sight at Great Yarmouth on 14 June before working 2C35, the 18.47 back to Norwich. The Vossloh worksplate can be seen under the loco number on the cabside; 68026 onwards featured Stadler worksplates after the takeover of the Spanish manufacturer. The later Stadler locos can also be identified by the absence of the round Vossloh logo between the headlights.

On its first day of operation with Greater Anglia and only the second service, 68027 is seen at the end of the line in Great Yarmouth on 14 June, having just worked the 2P34 18.04 from Norwich. Note the temporary plain blue livery which the loco wore before joining the TransPennine Express fleet, and the Stadler worksplate under the loco number.

One 'short set' meets another as 68005 *Defiant* and 68024 *Centaur* prepare to depart from Norwich with 2P34, the 18.04 to Great Yarmouth, while BR green 37057 and Colas 37116 prepare to follow shortly afterwards while working the 1Q18 14.14 Cambridge–Cambridge Network Rail test train on 20 July 2017. The test train is booked to cover many routes around East Anglia every four weeks, with the Thursday always featuring a trip to the Wherry Lines, bringing the welcome sight of a pair of Colas Class 37s to Norfolk. 37057 has since been repainted into the standard Colas colours with the headcode blinds and bufferbeam skirts removed.

A regular sight in Norwich while the second 'short set' was operating; after working the 10.05 Norwich to Lowestoft and return, the DRS Class 68s would stable in the Jubilee Carriage Sidings next to platform 6 until forming the 16.38 to Great Yarmouth, unless a DMU could not be sourced to cover the 12.05 and 14.05 departures to Lowestoft. Reasons may have been found to ensure that a unit was frequently unavailable for these two trips, necessitating extra mileage for the popular Class 68s! However, on 22 July 2017, a DMU had clearly been available, giving 68024 *Centaur* and 68005 *Defiant* a rest before the afternoon trips to Yarmouth.

For many years, Great Yarmouth was associated with loco-hauled services, but never with 'heavyweight' Class 37/7s, which were designed for freight services. It is therefore understandable the amount of interest generated when 37716 joined the small ranks of locos to operate the Greater Anglia 'short set', with the last passenger working for a Class 37/7 to Yarmouth being 37709 on 28 July 1990. The attractive Type 3 is pictured on the first day of operation to the Norfolk resort before working 2P33, the 18.17 to Norwich, on 25 August 2017. 37716 was hired to operator GIF for the construction of a high-speed line in Spain from 2001, but returned to the UK in 2013.

On the last day of operation for the Class 68s, 8 September 2017, 68001 *Evolution* rests on the rear of 2P28, the 16.38 service to Great Yarmouth, waiting for 68028 to do the hard work. The Class 68s had been a very popular part of the East Anglian railway scene while running on the Wherry Lines during 2016 and 2017, with the services on the last day understandably busier than usual. East Midlands Trains 158864 can be seen on the left preparing to work the 16.54 to Manchester Piccadilly.

37716 passes Berney Arms, one of the most inaccessible stations in Britain, as it has no road access, while working the 2C65 15.55 summer Saturday additional service from Great Yarmouth to Norwich, with 37405 hidden behind the trees on 9 September 2017. The station sign can just be seen on the right. The line from Great Yarmouth to Reedham closed for 16 months from October 2018 for resignalling, with all traffic routed on the more direct line via Acle. Owned by English Heritage, Berney Arms windmill is on the left, while Mutton's Mill in Halvergate can be seen in the distance, this being named after the last keeper, Fred Mutton, after it stopped being used as a windpump in 1947. The River Waveney can be seen in the foreground.

Under starters orders at Norwich as 68001 *Evolution* waits patiently for 90012 *Royal Anglian Regiment* to depart with the 13.30 to Liverpool Street on 16 September 2017, before following directly after at 13.33 with 1G03, the EACH Express 2 charter to Liverpool Street. The tour was organised by Greater Anglia to celebrate the end of Class 68 operation and followed a similar charter in April 2016 hauled by 37405 and 37419 in aid of East Anglia's Children's Hospices. The set had previously run to Ely and back, with 68034 at the other end of the stock; these locos being chosen as they were the oldest and newest members of the fleet.

The fine signal protecting Oulton Broad North Junction can be seen to the right of 37419 *Carl Haviland 1954-2012* and 37425 *Sir Robert McAlpine/Concrete Bob* working 2J83, the 15.48 Lowestoft–Norwich, on 23 March 2018. The East Suffolk Line on the right was singled in 1986 with Radio Electronic Token Block (RETB) signalling adopted, necessitating the fitting of signalling equipment to all locos or units which traversed the line. In 2012, the route was resignalled with a conventional track circuit block with axle counters system. At the same time, a passing loop in Beccles was installed, allowing an hourly service to commence, the long section of single track to Halesworth having previously only allowed a two-hourly service.

For once, 153314 is taking centre stage, with the ever-popular 'short set' on the right. The single-coach unit is arriving at Lowestoft with the 14.08 from Ipswich on 18 May 2018, which connects with the 2J83 15.48 departure to Norwich, to be powered by 37419 *Carl Haviland 1954-2012* and 37407. The fleet of five Greater Anglia Class 153s moved to Transport for Wales in December 2019, being replaced by the new and much more spacious Stadler bi-mode units.

After arrival into Norwich, 37419 *Carl Haviland 1954-2012* catches the sun before continuing with the regular diagram to form 2P32, the 17.36 to Great Yarmouth, on 18 May 2018. 37407, seen on the far end of the stock, had been purchased by DRS from the Churnet Valley Railway like 37424 before being the subject of a thorough and expensive overhaul, with its first passenger service for 18 years taking place with Greater Anglia on 19 February 2018.

On the first day of passenger services with Greater Anglia, 28 June 2018, newly overhauled 37409 *Lord Hinton*, named after the British nuclear engineer who supervised construction of the first large-scale nuclear power station, looks immaculate at Great Yarmouth before working the 2P33 18.17 to Norwich. As part of the overhaul, 37409 was fitted with DRA and remote fire extinguisher equipment, both being essential safety features specified by Greater Anglia. This was also the first day where two BR large logo Class 37s worked the 'short set' with 37407 visible on the rear.

Making a fine sight storming away from the booked station stop, recently refurbished 37409 *Lord Hinton* powers away from Brundall while working the 2P33 18.17 Great Yarmouth-Norwich service on 29 June 2018, with classmate 37407 helping at the rear. Before its overhaul, 37409 had been a regular performer on the Cumbrian Coast line for Northern but this was only the second day of operation for Greater Anglia. Until the mid-1960s, a small row of railway cottages could be found adjacent to Brundall station building, behind the bushes on the left of the picture.

The combination of the first pair of large logo Class 37s on the Greater Anglia 'short set', newly refurbished 37409 *Lord Hinton*, and several days of full sun proved to be irresistible to most linesiders! 37407 leads 37409 through Postwick on the approach to Brundall Gardens, while working 2P32, the 17.36 Norwich–Yarmouth service, on 2 July 2018. For many years, the bridge just visible on the right featured a 20mph speed restriction for heavier axle loads, such as loco-hauled services; it was finally replaced in January 2021, which allowed the speed restriction to be lifted.

Although the mechanical signalling can still be seen at Acle on 2 July 2018, the modern equipment was gradually changing the scene by this date, with a large radio mast casting a long shadow. Despite this, the sight of a large logo Class 37 on a passenger train pausing at a rural station was quite incredible in 2018! 37409 *Lord Hinton* leads 37407 away from Acle working the 2P33 18.17 Great Yarmouth–Norwich service. The photograph was taken from a road bridge over the station which was constructed in the late 1980s.

Although all the Class 37s were well over 50 years old by 2018, it was often the 1970s Mk.2 coaches which caused issues. This was the case on 19 July, when only two coaches were available for service, producing what became known as the 'shorter set'! Even when all coaches were available for service, the air-conditioning rarely worked by this date, with DRS reluctant to invest in rolling stock that had a very limited future. 37419 *Carl Haviland 1954-2012* is pictured leading 37405 through Postwick towards Brundall Gardens, with the 2J88 19.02 Norwich–Lowestoft.

A virtually unspoilt country junction, featuring mechanical signalling and a large logo engine, as 37407 eases off the Berney Arms line at Reedham on 11 August 2018 while working 2C61, the 13.55 Great Yarmouth–Norwich, with 37423 *Spirit of the Lakes* on the rear. The ensemble paused under the bridge for some time, believed to be related to a slipping fault on 37423 when driven remotely from the leading loco. Note the speed limit on the right for the sharply curved line round to the swing bridge and on towards Lowestoft.

The headlight on 37424 *Avro Vulcan XH558* sparkles as it leads 37405 into Norwich with the 2P13 09.17 from Great Yarmouth on 27 August 2018. The centre road between platforms 4 and 5 was once used for run round manoeuvres, but the pointwork is now disconnected, with the track only used to stable locos and stock. A set of Greater Anglia Mk.3 coaches can be seen on the right, with a service for London Liverpool Street.

Right: Later on 27 August, it is 37405's turn to lead with 2J80, the 14.55 Norwich–Lowestoft, seen here snaking off Somerleyton swing bridge towards Somerleyton station, with 37424 *Avro Vulcan XH558* on the rear. Like the similar bridge in Reedham, the current structure in Somerleyton (just beyond the picture on the left) replaced the original single line bridge in 1904. Until 1965, a signal box on the Norwich-bound platform controlled the station but after closure, control of the station signals was transferred to the swing bridge box. Signalling for the whole line is now managed remotely from Colchester, although the box still controls the opening of the swing bridge to river traffic.

Below: On 31 October 2018, with headlights blazing, 37419 *Carl Haviland 1954-2012* prepares to lead 37716 away from Norwich with the 2J88 1902 Norwich–Lowestoft. Depending on the rostered driver, departures could, at times, be quite spectacular. This resulted in most 'short set' departures being booked from platform 6 towards the end of operations, to avoid disturbances in the main trainshed, as the highest numbered platform was furthest from the main concourse.

The old and the new: the semaphore signals were still in use on 22 December 2018 at Lowestoft, although the new colour light signals had been installed ready for commissioning as part of the £68m project to resignal the Wherry Lines. A regular performer on the 'short set' since 2015, 37405 rolls into the most easterly station in Great Britain with 2J80, the 14.55 from Norwich, with 37423 *Spirit of the Lakes* helping on the rear. Unlike most other DRS Class 37/4s, 37405 was not refurbished and, therefore, by 2021, it was held as a strategic reserve, with parts being donated to keep others in the fleet running.

The guard closes the door on the brake van, ready for the departure of the 2J83 15.48 Greater Anglia service to Norwich at Lowestoft, with 37423 *Spirit of the Lakes* ready to provide the power, with assistance from 37403 *Isle of Mull* on the rear, on 24 December 2018. By this stage, the 'short set' was clearly on borrowed time, although the end date had not been announced; every extra month of operation was a bonus.

Christmas Eve in Norwich. Never mind the seasonal decorations, the sight of a preserved Class 37 working regular passenger services is much more impressive! The station lights on platform 6 illuminate 37403 *Isle of Mull* after working 2J83, the 15.48 from Lowestoft. The impressive English Electric loco continued to work for Greater Anglia until 2 February 2019, with a return to Bo'ness in August 2020, after the end of the DRS hire contract. On return to Scotland, the Scottish Railway Preservation Society installed a new engine in the loco.

Who could have predicted a 'no heat' Class 37 working a passenger service from Lowestoft in 2019! 'Heavyweight' 37716 pauses before hauling 2J83, the 15.48 to Norwich, on 21 March, with assistance from 37407 at the rear. Until the late 1960s, the wooden station roof extended onto both platforms. In 1992, the remaining roof over the concourse was removed, despite much local opposition, although it was deemed necessary given its poor state of repair. The Lowestoft Central Project, which has already made vast improvements to the station in recent years, has aspirations to install a new roof over the concourse.

On 18 May 2019, Greater Anglia ran the EACH Express 3 charter from Norwich over routes associated with the Class 37s as a farewell to the 'short set'. At this stage, it was not known exactly how much longer the set would work, but it was clear the end was in sight. 37405 pauses at Ely accompanied by 37409 *Lord Hinton*, before reversing to head to King's Lynn as 1G38 at 11.35. Although this was new ground for the Greater Anglia set, Class 37s had regularly worked between Liverpool Street and King's Lynn in the past before the line was electrified, when services to the capital transferred to King's Cross.

Once a regular sight in Liverpool Street, Class 37s (and in fact, any diesel locos) are now a rarity at this terminus. 37409 *Lord Hinton* pauses after running from King's Lynn via the West Anglia line, and before returning to Norwich via the Great Eastern Main Line. In the preceding days, it had been hoped that 37402 *Stephen Middlemore 23.12.1954-8.6.2013*, one of only two DRS Class 37/4s not to have worked the 'short set', may have featured on the charter but the two regular locos, 37405 and 37409, were chosen on the day. Note the slightly longer set of five coaches; despite this, tickets for the tour sold out quickly.

37405 poses at Liverpool Street, providing a change from the regular electric traction after arrival from King's Lynn. The terminus was redeveloped in the late 1980s, resulting in a very different setting compared to the days when Class 37s were regularly used on the main line to Norwich. The EACH Express 3 charter raised £22,000 for East Anglia's Children's Hospices, which provide support for families and care for children and young people with life-threatening conditions, with a total of £54,000 raised from the three charters in 2016, 2017 and 2019.

Friday 26 July 2019 was a 'will it, won't it' day with the first part of the regular 'short set' diagram covered by DMUs. 2P20, the 12.36 to Great Yarmouth and return, featured 37716 and 37407, but the 2C63 14.55 was again covered by a unit. Therefore, the sight of 37716 shunting into the platform for the 2P32 17.36 to Yarmouth was a welcome one, even if the set had returned to 'shorter set' formation with only two coaches available. The warmer summer weather and lack of opening windows did not help the internal conditions of the DRS Mk.2 coaches, with the air-conditioning rarely working by this date.

The appearance of the first DRS Class 37/4 in BR large logo livery took some by surprise, as previously all traction from the Cumbrian-based company had worn its corporate livery. However, the variation was taken even further with the appearance of 37419 *Carl Haviland 1954-2012* in InterCity Mainline livery in 2019. The Type 3 definitely catches the eye at Lowestoft before working 2J85, the 16.48 to Norwich, a service which usually employed a Greater Anglia DMU, on 1 August 2019. Large logo 37407, by now named *Blackpool Tower*, was on the rear. The livery variations kept coming with 37425 *Sir Robert McAlpine/Concrete Bob* repainted into Regional Railways livery after overhaul but, sadly, this happened too late for it to appear on the 'short set', although it was a very popular sight on Railhead Treatment Trains in the Anglia region in 2020.

The old and the new as 37424 *Avro Vulcan XH558* and 37407 *Blackpool Tower* prepare to depart from Norwich on 31 August 2019, with 2P32, the 17.36 to Great Yarmouth, while Stadler Flirt bi-mode unit 755410 rests in the centre road. The 'short set', plus the entire DMU fleet based at Crown Point depot in Norwich, comprising Classes 153, 156 and 170, were replaced by these brand-new units. Although they did not have the character of the 1960s engines, the comfort and space for passengers was a big improvement on the Mk.2 coaches, which had definitely seen better days!

After many months of anticipation, the final day of operation of the Greater Anglia 'short set', 21 September 2019, was advertised in advance, enabling enthusiasts and photographers to enjoy the sight and sound of English Electric traction on the Wherry Lines for one last time. Large logo pair, 37424 *Avro Vulcan XH558* and 37409 *Lord Hinton*, had the honour of working on the final day, and are pictured pausing at Brundall while working the 2P20 12.36 Norwich–Great Yarmouth. The atmosphere on-board on this day was very good, with plenty of enthusiasts out for one last ride as well as interest from regular passengers.

The classic shot from the A143 road bridge at Haddiscoe, with the passengers on 2J81, the 14.57 Lowestoft–Norwich, clearly appreciating the last day of loco haulage, in the form of 37409 *Lord Hinton* and 37424 *Avro Vulcan XH558*. The rostered drivers and guards also helped to make this a very enjoyable day for everyone, with excellent weather an added bonus. Haddiscoe was once a junction with the Yarmouth Southtown to Beccles line, with a spur just beyond the large tree on the right connecting to the East Suffolk Line via Fleet Junction.

Transition time: the stop boards on the left at Haddiscoe show stopping positions for the 'short set', pictured on the final afternoon of operation with 37424 *Avro Vulcan XH558* and 37409 *Lord Hinton* working 2J82, the 15.50 Norwich–Lowestoft, as well as signs for the new Class 755 units which have now replaced them. The loco-hauled operation was given a fine send off by Greater Anglia, with a full day of English Electric action, and an extra round trip to Lowestoft compared to the usual diagram. Few would have guessed back in July 1994, when 37077 hauled two trips to Great Yarmouth and one to Lowestoft, that the final Class 37-hauled service on the Wherry Lines would be 25 years later on 21 September 2019, with 37409 hauling the 2P39 20.17 Great Yarmouth–Norwich.